P9-DEY-738

The Galápagos Tortoise

A Dillon Remarkable Animals Book

The Galápagos Tortoise

By Susan Schafer

DILLON PRESS
New York

Maxwell Macmillan Canada
Toronto

Maxwell Macmillan International
New York Oxford Singapore Sydney

Acknowledgments

My sincerest gratitude goes to Peter Pritchard, whose books on turtles have always been an invaluable source of information, for taking time to review this manuscript. I owe my thanks to Val Lance and Charlie Radcliffe of the San Diego Zoo, to Jim Murphy of the Dallas Zoo, and to my friend and editor, Joyce Stanton, for her hard work and encouragement.

Photo Credits

All photos courtesy of Susan Schafer

Painting on page 18 courtesy of the American Museum of Natural History, #990.

Library of Congress Cataloging-in-Publication Data

Schafer, Susan.
The Galápagos tortoise / by Susan Schafer. — 1st ed.
p. cm. — (Remarkable animals series)
Includes bibliographical references (p.) and index.
Summary: Describes the physical characteristics, ancestry, habits, and habitat of this living fossil.
ISBN 0-87518-544-4
1. Galápagos tortoise—Juvenile literature. [1. Galápagos tortoise. 2. Turtles.] I. Title. II. Series.
QL666.C584S32 1992 597.92—dc20 92-7396

Dillon Press
Macmillan Publishing Company
866 Third Avenue
New York, NY 10022

Maxwell Macmillan Canada, Inc.
1200 Eglinton Avenue East
Suite 200
Don Mills, Ontario M3C 3N1

Macmillan Publishing Company is part of the Maxwell Communications Group of Companies.

First edition

Printed in the United States of America

10 9 8 7 6 5 4 3 2 1

Contents

Facts about the Galápagos Tortoise

Scientific Name: *Geochelone elephantopus* is the name used most often*

Description:

Length—Curved measurement of top of shell: rarely over 5½ feet (1.7 meters), depending upon the island; domed males average about 4½ feet (1.4 meters), females about 3 feet (.9 meter); saddlebacked males measure about 3 feet (.9 meter), females about 2 feet (.6 meter)

Weight—Up to 681 pounds (309 kilograms), depending upon the island; domed males normally weigh about 500 pounds (227 kilograms), females about 250 pounds (113 kilograms); saddlebacked males weigh from 100 to 150 pounds (45 to 68 kilograms), females about 50 pounds (23 kilograms)

Physical Features—Small head, long neck, leathery skin; heavy elephantlike feet and legs, thick toenails; large, rounded shell; males have long tails, and females have short tails

Color—Gray brown, some with pale yellow patches on face

Distinctive Habits: Herbivore, browser, grazer; often found in groups; on some islands, migrates to moist highland areas during the dry season

Food: Grasses, herbs, leaves; sometimes fruit, flowers, and fallen cactus pads

*Many names have been given to the Galápagos Tortoise, but scientists have not agreed on one.

Reproductive Cycle: Mature in about 25 years; females normally lay 5 to 9 round, hard-shelled eggs several times a year; eggs take 3 to 6 months to hatch, depending upon the temperature; young are able to take care of themselves immediately after hatching

Life Span: Long-lived, probably more than 100 years

Range: Found only on the Galápagos Islands, located about 600 miles (968 kilometers) off the coast of Ecuador in South America; to 5,600 feet (1,707 meters) in elevation

Habitat: Dry lowland forests, open grasslands, damp highland forests

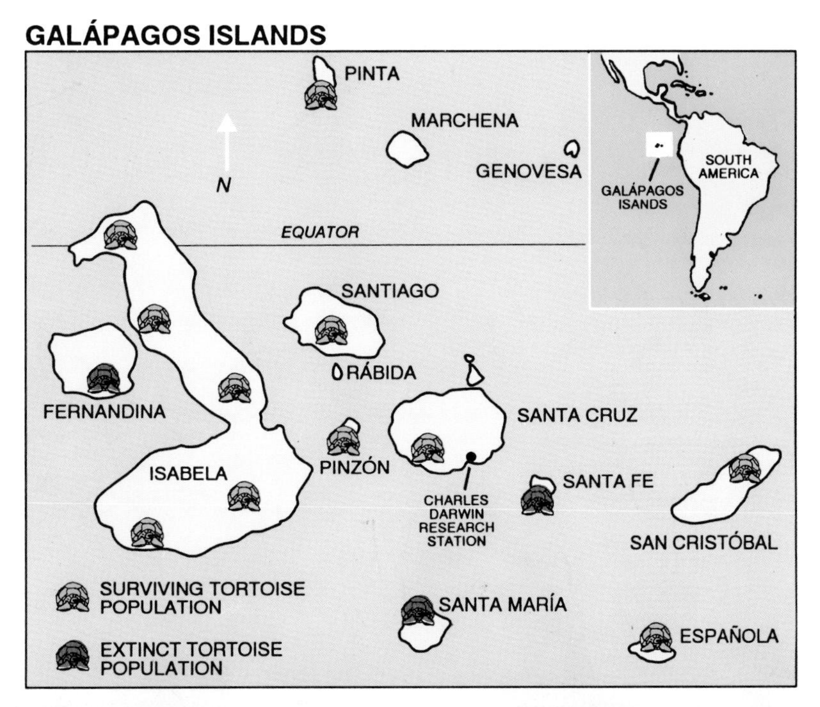

Chapter 1

The World's Largest Tortoise

A thick fog sweeps across a grassy clearing atop the rim of an old volcano, blanketing the ground with shimmering drops of dew. A small geyser hisses softly, spurting steam into the air. A smooth, dark boulder glistens in the mist like a sorcerer's caldron turned upside down.

As if a spell had been cast, the boulder lifts from the ground, floats forward, and settles back into the grass. A brown hawk with a spotted chest, startled by the movement, flutters into the air and circles away into the haze. For a moment, all is still.

Then, from beneath the boulder, a gray head with dark eyes and a jagged beak rises on a long, leathery neck. From the beak, stiff blades of grass protrude like green toothpicks. Lifted by four mas-

Seemingly lost in a magical world, the Galápagos tortoise lives only on a group of rugged, volcanic islands off the coast of South America.

With its heavy shell and huge body, the Galápagos tortoise is the largest tortoise living in the world today.

sive gray legs, the boulder moves off once again, rocking from side to side. The boulder, of course, is the shell on the back of the Galápagos (guh-LAH-puh-gus) tortoise, the largest living tortoise on earth.

Breaking the Record

Supposedly, the largest living tortoise is the Aldabra tortoise from the Aldabra Islands in the Indian Ocean near Madagascar. The largest Aldabra weighed 657 pounds (298 kilograms). However, an old male Galápagos tortoise, named Number 5, lives at the San Diego Zoo in California. At his peak, he weighed 681 pounds (309 kilograms), more than two refrigerators put together. That breaks the Aldabra record by 24 pounds (11 kilograms)!

There aren't many weight records for Galápagos tortoises in the wild. Can you imagine carrying a big-enough scale to the top of a volcano? Most adult Galápagos tortoises, however, probably weigh less than 500 pounds (227 kilograms). Although the Galápagos is the largest tortoise, it is not the largest turtle. A tortoise is a turtle that lives only on land. The leatherback sea turtle may weigh up to 1,900 pounds (861 kilograms), more than three times the weight of most Galápagos tortoises.

Tortoise Islands

The Galápagos tortoise lives only on the Galápagos Islands, located in the Pacific Ocean about 600 miles (968 kilometers) off the coast of Ecuador, in South

America. *Galápago* is the Spanish word for tortoise. When you say "Galápagos tortoise," you are saying "tortoise" twice.

The islands have had many names since they were discovered in 1535 by Fray Tomas deBerlanga, the bishop of Panama. Tomas's ship had been caught in calm seas and swept along by strong ocean currents to a land where strange, wild animals had no fear of people.

Birds flew aboard his ship, boldly poking into his gear and stealing bits of food. Hundreds of fat, spiny-headed iguanas lounged fearlessly along the shore. Tortoises of gigantic proportions, larger than any the crew had ever seen, grazed on the hills as placidly as cows. Because the islands seemed to have magically appeared out of the ocean mists—and because of the tameness of the animals—they were first called *Las Islas Encantadas*, or the Enchanted Islands.

The animals on the Galápagos were tame because there were no large **predators***, like wolves or jaguars, to threaten them. They were safe on their

*Words in **bold type** are explained in the glossary at the end of this book.

The Galápagos Islands: a rocky, moonlike land of hills and craters

islands and were not afraid to approach other creatures, like the sailors on Tomas's ship. Living on islands also affected the way the animals developed physically. Tortoises grew larger and larger over time, for example, because they had no predators and no other competitors for food.

Animals that live on islands are often very different from those that live on continents. They live in smaller groups, isolated from larger populations.

Year after year, generation after generation, they mate among themselves. Only those animals with characteristics that help them survive in their particular island environment live to mate again. Year after year, they become better **adapted** to their surroundings. Eventually, over thousands of years, they develop into unique—or one-of-a-kind—**species,** like those discovered on the Galápagos.

Of all the animals on the Galápagos, the giant tortoises impressed the mariners most. Soon, in spite of the original name, the islands became known as *Las Islas Galápagos,* or the Tortoise Islands.

Thirteen major islands, and scores of islets and rocks, form the Galápagos group. Originally, many of the islands were active volcanoes. Today, only a few of the younger volcanoes at the western end of the chain are still active. At the eastern end, closer to South America, scientists recently discovered ancient, sunken islands that were once part of the Galápagos.

The equator runs right through the top of the

largest Galápagos island, named Isabela after a Spanish queen (or Albemarle after an English duke). Over the years, each island has collected an assortment of official Spanish and English names.

Older than People

Giant tortoises live longer than any other land animal, possibly for as long as 200 years. Scientists aren't sure because the tortoises outlive the people who keep track of them, and there are few records.

The oldest recorded age for a giant tortoise is 152 years. It was for a tortoise that lived on a British military post in the Mascarene Islands, east of Madagascar. One of the oldest Galápagos tortoises is Number 5, at the San Diego Zoo. When he came to the zoo in 1933, he weighed 475 pounds (215 kilograms). An adult tortoise that size normally would be about 40 or 50 years old. Add that to the 59 years he has been at the zoo, and he would be at least 99 years old, if not well over 100. And he's still going strong!

Why do Galápagos tortoises live so long? Scien-

Reaching ages of well over a hundred years, giant tortoises easily outlive the people who study them.

tists aren't really sure. Sometimes, however, larger animals live longer than smaller ones. For example, the largest land mammal, the elephant, weighs more than 10,000 pounds (4,536 kilograms) and lives more than 70 years. The smallest land mammal, the pygmy shrew, weighs less than 1/10 of an ounce (2.9 grams), about the weight of an almond without its shell. It lives barely more than a year.

A Turtle or a Tortoise?

Turtles are reptiles that have shells. They belong to an **order**, or a large group of animals, called *Chelonia*. The order is divided into 12 **families**, or smaller groups, of different kinds of turtles. One of these is the tortoise family, or *Testudinidae*. Tortoises are turtles that live on land.

The use of names like turtle and tortoise can be confusing. Part of the problem starts when people from different countries call turtles by different names. In Australia, only sea turtles are called turtles. All other turtles there are called tortoises even though they live in rivers and ponds.

That's why scientists use **scientific names**. Every animal is supposed to have only one scientific name. Then, it cannot be mistaken with any other animal. There is a problem, however, when scientists don't agree upon a correct scientific name. The Galápagos tortoise, for example, has been given many names since it was first called *Testudo nigra*—which means "black tortoise" in Latin—in 1824. Today, most scientists use the name *Geochelone elephantopus*, which means "land tortoise with the elephant feet" in Greek. Whatever you call it, one thing is certain—most people like the tortoise better than any other reptile.

Today's turtles have changed little from these prehistoric ancestors.

Chapter 2 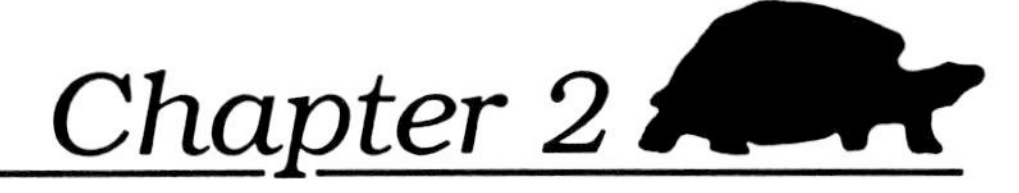

Life in a Shell

Imagine a time on earth more than 200 million years ago. Before there were mammals and birds. Before there were crocodiles, lizards, and snakes. Picture the dinosaurs tramping through strange, primitive forests, unlike any today, and wading through dark swamps.

Look closer and, along with the dinosaurs, you will see turtles swimming in ancient waters. They cannot pull their heads into their shells, and they have teeth, but they are undeniably turtles. They are the ancestors of the turtles we know today.

Living Fossils

Turtles are the only **vertebrates** with a hard, solid shell. The shell is made of bone and, along with other bones in the body, can slowly harden in the earth to form a **fossil**. By studying fossils, scientists have discovered that turtles have not changed much since those prehistoric days, except they now lack teeth and most can pull in their heads. Because they have changed so little, they are sometimes called "living fossils."

Turtles, with their heavy shells, cannot jump, sprint, or fly, but that hasn't stopped them from getting about. Some 250 species live around the world on every continent except Antarctica. Is it hard to imagine turtles in the snow? They may not live in the Antarctic, but many live in climates with freezing winters.

Frozen but Alive

In North America, the painted turtle spends the winter underground with part of its body water frozen solid. It doesn't breathe, and its heart doesn't beat. When spring warms the earth, the turtle thaws and leaves its winter retreat.

From the icy north to the tropics, turtles live in many **habitats**. Some species, like the North American snapping turtle and the Australian snake-necked turtle, live in ponds and rivers. Sea turtles live only in the ocean. Tortoises live in forests, grasslands, and deserts.

Different Looks for Different Life-Styles

Turtles look different depending upon where they live and how they manage to survive. Sea turtles have flipperlike legs and sleek shells, like the hulls of racing boats, because they swim in the ocean. Racing through the water for short distances at 20 miles (32 kilometers) per hour, they are the fastest kind of turtle.

Tortoises have stout legs and feet because they walk on land. Their shells slow them down. The Galápagos tortoise travels only 4 miles (6.4 kilometers) a day, with plenty of rest stops. The tortoise would take almost a week to go as far as the

Tortoises have sturdy legs and feet for walking on land.

sea turtle goes in one hour.

Pond turtles have characteristics of both land and sea turtles because they live in the water and on the land. Some, like the North American box turtle, have hinged shells that open and close like a clam's. When frightened, they shut themselves inside.

A Box of Ribs

In spite of their differences, all turtles have a shell. They can't crawl out of their shell like a hermit crab. It is a permanent part of their bodies. Depending

The turtle's shell is made of flat bony pieces that fit together like a puzzle.

upon the species, the shell has about 60 bones.

You have similar ingredients for a shell—you're just put together differently. Reach down and feel your ribs. They form a rib cage around your **internal** organs, like your heart and lungs. Imagine your rib cage growing bigger until it was big enough for you to pull your arms, legs, and head inside. You'd almost have a turtle's shell.

On a turtle's shell, however, the spaces between the ribs are filled with platelike pieces of bone that fit together like a puzzle. The plates are attached to one another, and to the ribs, spinal column, shoulder bones, and hip bones. The result is a solid, protective box that makes up as much as one-third of the animal's total weight.

Like a box, the shell has two parts. The upper part is called the **carapace**, and the lower part is called the **plastron**. Thin, horny shields, called **scutes**, fit like puzzle pieces over the bone plates, just as fingernails cover the ends of your fingers. Scratch the top of one of your fingernails. You can feel what a turtle feels when something scratches its shell.

Armored Shell

The shell is like a suit of armor. It protects the turtle from the crushing jaws of predators. When threatened, many turtles simply pull in their heads and legs and wait until danger passes.

The shell protects the turtle in other ways, too. In

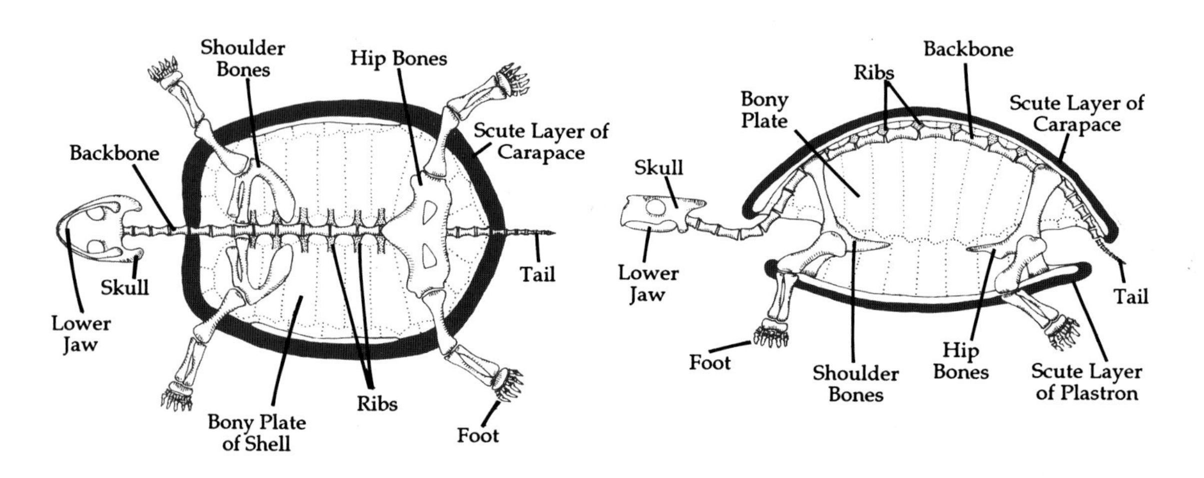

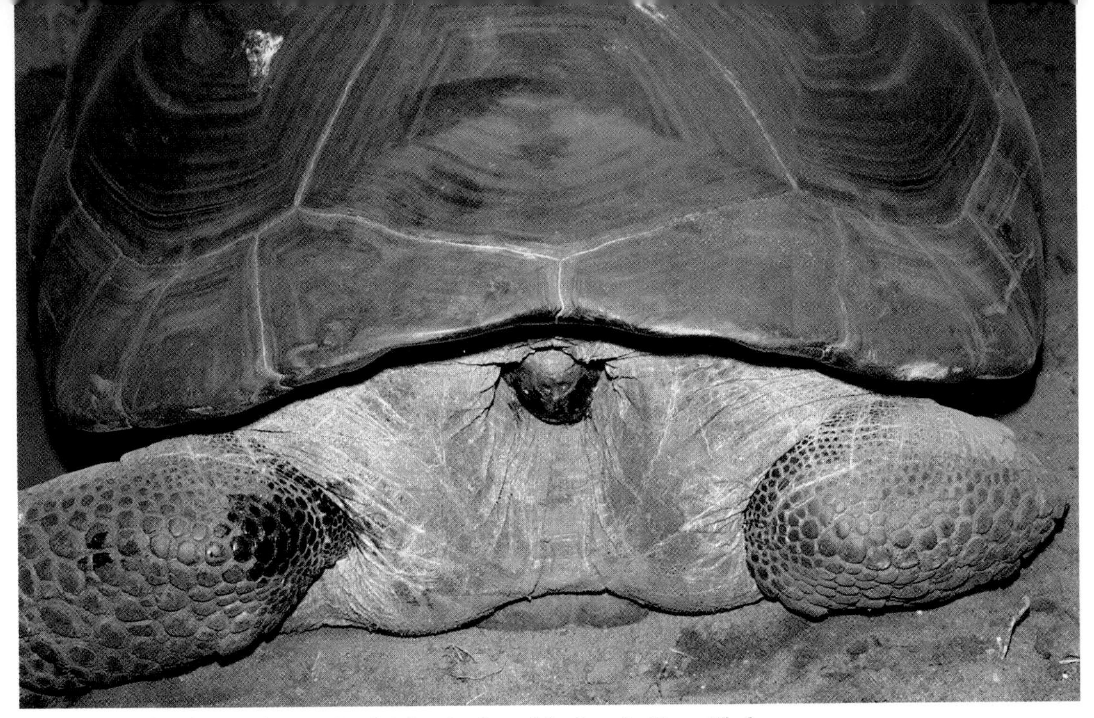

A frightened tortoise hides its head in its shell until danger passes.

places like the desert, where water is precious, a turtle might **dehydrate** if it lost moisture from its body. Its shell then works like a padded canteen, preventing the animal from drying out.

The shell also makes a good heater. Have you noticed how quickly a blacktop road heats up in the summer sun? A turtle's dark shell absorbs heat the same way. Once it is warm, the shell becomes an **insulator**, like a cozy jacket, to keep the turtle warm. If it gets hot, the turtle moves into the shade to cool. Like other reptiles, the turtle is cold-blooded; it cannot make its own body heat. Instead, its temperature matches its environment.

The Tortoise Family

Tortoises live in the warm **tropics** and **subtropics** of every continent except Australia and Antarctica, although most live in Africa. While the Galápagos tortoise is the largest, the smallest is the speckled tortoise of Africa. Its shell is never more than 4 inches (10 centimeters) long—not much bigger than the egg of the Galápagos tortoise.

Because all tortoises are **terrestrial**, they are sometimes called land turtles. They live a slow, easy life, **browsing** on grasses, leaves, flowers, and fruit. Like the Galápagos tortoise, they are **herbivores**. Some tortoises eat insects and small **invertebrates**, like earthworms, millipedes, snails, and slugs, when they can catch them. Most will eat carrion, the flesh of dead animals, if they find it.

How Smart Is a Tortoise?

Tortoises have sharp eyes and intelligent minds. Turning their long necks like periscopes on a submarine, they can look in every direction. As curious as

cats, they investigate anything new, especially red and yellow objects, because they are the colors of flowers and fruit.

Some scientists wanted to find out if Galápagos tortoises could see colors other than red or yellow. They conducted an experiment that not only answered their question about the tortoise's vision, but also gave them a clue to the animal's intelligence.

The scientists showed the tortoises red, yellow, blue, and green lights. If the tortoises went to the blue and green lights, they received a food reward. If they went to the red and yellow lights, they got nothing. The tortoises quickly learned to go to the blue and green lights.

About a year later, the tortoises were again shown the colored lights. Every one remembered to go straight to the blue and green lights. Not only can they see many colors, but they also have excellent memories. Perhaps it's the tortoise, not the elephant, that never forgets.

Chapter 3

Tortoise Islands

Craaash! Wooosh! Craaash! Wooosh! Wave after wave rolls in from the sea and smashes against the rocky coastline of a Galápagos island. Spiny marine iguanas, cold from grazing on underwater seaweed, wash ashore and scramble up the rocks to warm themselves in the sun. A sleek sea lion barks irritably.

Inland, ancient lava flows stretch for miles like asphalt freeways. Blocks of pitted rock, once spit like seeds from the mouth of an erupting volcano, litter the ground between stunted bushes and thorny trees. Treelike prickly pear cactuses spread their fat leaf pads under the hot, equatorial sun.

Crowded beneath a giant prickly pear, several Galápagos tortoises nap in the afternoon shade. The season is changing, drying the flowers, grasses, and

Marine iguanas sun themselves on one of many hardened lava flows that cover the Galápagos Islands.

leaves. Soon, the tortoises will move to higher elevations, where green grasslands thrive among the trees.

High and Wet or Low and Dry

Not every island in Galápagos, however, offers a rich highland living for the tortoises. Only some of the larger islands, like Isabela and Santa Cruz, have peaks high enough—over 2,000 feet (610 meters)—to trap the clouds and moisture.

Small islands, like Española and Pinzon, that reach only a few miles across and a few hundred feet high, are hot and dry. Tortoises must compete for limited food, shade, and places to sleep. Because of these different conditions, tortoises on smaller islands look different from those on larger islands.

Domes and Saddlebacks

The tortoises on the largest islands are called domes. They have large shells that look like the roof on your state capitol. Males can weigh more than 600 pounds (272 kilograms). Tortoises on smaller islands do not

Tortoises that live on high, wet islands have large shells and are called domes. Here, a female dome settles into her bed for the night.

grow so large. The largest males there rarely weigh more than 150 pounds (68 kilograms). Because their shells are stretched up in front like a horse's saddle, they are called saddlebacks.

Each shell type is adapted to its own environment. Domes live in wetter, cooler environments where their large shells hold in more heat. A big body takes longer to heat up, but once it is warm, it holds

Tortoises that live on low, dry islands in the Galápagos have small shells that are raised in front like a horse's saddle.

the heat longer. Saddlebacks live in drier, hotter environments. Their small shells help to keep them from overheating. A small body is quicker to heat up, but it is also quicker to cool off.

The passing of time has molded the different shapes of tortoises on the different islands. The earliest tortoises would have looked alike, and before that, they weren't there at all.

Island Pioneers

More than nine million years ago, the Galápagos Islands did not exist. Then, in a burst of volcanic activity, they sprang from a crack in the bottom of the Pacific Ocean. Molten lava poured into the sea, building layer upon layer of rock on the ocean floor. Eventually, the rising land broke through the ocean's surface, and the Galápagos Islands were born.

At first, the islands were **barren**. Then, ever so slowly, seeds and spores blew on the winds from South America to sprout in the volcanic soil. Insects, birds, and sea lions, swept by winds or ocean currents, flew or swam to the islands. Some birds carried seeds or insects attached to their feet or feathers.

The tortoise, however, couldn't swim, fly, or hitch a ride on a bird. How did it get hundreds of miles across the ocean? It floated! As buoyant as a cork, the tortoise's shell acted like the hull of a boat to keep it afloat. The tortoise may also have managed to crawl onto a natural raft of logs, branches, and vegetation. Drifting on the currents, it could survive for months without food or water—its **metabolic rate** is so slow that it wouldn't need to eat. Scientists believe that just one female tortoise carrying eggs would have been enough to start a colony of tortoises in the new land.

Nothing to Fear

The first tortoises in the Galápagos were probably smaller than the ones we know today. That is because their ancestors on the mainland needed to hide from predators and compete with other herbivores, like deer, for food. On the islands, however, they had nothing to fear and no other competitors. Over time, the species became the giants that we know today.

Grazing on grasses or browsing on leaves throughout the day, the tortoise clips plants as efficiently as a pair of garden shears. Taking a few bites, it takes a few steps, and then a few more bites. It may not have teeth, but its sharp beak and muscular jaws can break bone.

It takes a lot of food to fill a Galápagos tortoise. One could easily down 50 heads of lettuce in a single day. If you ate a large salad for breakfast, lunch, and dinner every day, it would take you more than three months to eat as much! Lettuce, however, would not make a nutritious diet for the tortoise. It prefers the tender new shoots of grasses and leaves.

Dry and Cool or Wet and Warm

With the arrival of the dry season, the lowland plants turn yellow. Then the tortoises munch on dried grasses and leaves. They may even eat the thick pads of prickly pear that have fallen to the ground, sharp spines and all!

The dry season lasts from about June or July through December. It rarely rains, but it isn't hot. Instead, temperatures drop into the 70s, and the islands are blanketed with cool, misty fogs called *garuas* (gah-ROO-ahs), from the Spanish word for "drizzles." Sunny days are about as scarce as green grass.

The wet season lasts from January until the start of the dry season. It often rains, turning the country-side green, but it isn't cold. Temperatures rise into the 80s. Between storms, the skies turn blue and the tortoises enjoy long spells of warm, sunny weather.

Parasite Patrol

Scruunch! Scruunch! In a forest clearing a Galápagos tortoise munches peacefully on tufts of grass. Sud-denly, a small female ground finch lands in front of the tortoise and hops about in the grass. The tortoise stops feeding and slowly rises, stretching out its neck and legs as far as they will go. Like a statue, it stands frozen in place.

Immediately, the bird hops onto the tortoise's leg, skitters up to its neck, and pecks at its skin. Soon, a male ground finch, as black as coal, joins the female, hopping onto the tortoise's head and pecking around its eyes.

Always on patrol for a meal, the finches keep the tortoise free of **parasites** by eating the ticks that live

A small finch, perched on the front leg of an outstretched tortoise, picks ticks off the reptile's leathery skin.

Sunk to the chin in mud, a Galápagos tortoise escapes the heat of the afternoon sun.

into the mud, forming a clutter of potholes that soon fill with rain. The water won't last long, however, because it sinks quickly into the loose, volcanic soil. Sticking their faces into their own footprints and sucking up the water, the tortoises drink while they can.

When large puddles form, the tortoises are sure to be in them. Like some other animals, tortoises like to wallow in the mud. The mud protects them from pesky insects, like mosquitoes, and keeps them warm on cool nights and cool on hot days. With a gurgle and a blurp, they sink blissfully into the muck.

Chapter 4

Who's the Tallest?

An old, saddlebacked male relaxes in the morning sun. Suddenly, a movement in the distance catches his eye. Stretching his neck to get a better look, he sees another tortoise enter his **territory**. Immediately, he races to meet the intruder, although a person at a slow walk could have beat him. The young intruder, surprised by the attack, stops in his tracks.

Facing each other, the tortoises lift their shells off the ground and stretch their necks high. With open mouths, they manage to look as fierce as possible. The object of their game is to be the tallest and fiercest without actually coming to blows. Like sports competitions for people, these **rituals** allow them to compete peacefully.

After a few minutes, the old male lifts his head a

Lifting his head high for a better view, an alert male saddleback spies an intruder in his terrritory.

Gaping fiercely, a victorious tortoise lifts his head higher than his opponent's.

few inches higher than his rival. The youngster quickly lowers his head, hissing loudly and tucking in his legs. With a thud, his shell crashes to the rocky ground. Turning abruptly, he leaves the area.

Winner Takes All

For a Galápagos tortoise, the higher it can lift its head, the higher it will be in the dominance **hierarchy**. It

will have the best feeding sites, sleeping sites, and mates. This is especially important on small islands where resources, like food, may be limited. Although saddlebacks are smaller than domes, they can reach higher because of their shape. The raised front of their shell does not block their neck.

Domes also stretch their necks and form hierarchies, but they don't quarrel as much as saddlebacks. They live in environments—islands with wet, green highlands—where they don't need to compete as much, except during the mating season.

The Chase and the Catch

In a highland forest, a male tortoise, his head stretched high in victory, watches an opponent crash off through the brush. Turning slowly, he stares at the female tortoise sitting nearby. It is early in the wet season and time for him to mate, but he's not sure the female will cooperate. He may have to spend an hour or two courting her before she will accept him. Cautiously, he takes a few steps in her direction. Instantly,

After a brief courtship, tortoises mate.

she moves away, and the chase is on.

Running as fast as his tortoise legs will carry him, the male struggles to catch the female. Closing in, he pulls in his head and lunges forward, smashing the front of his shell into hers.

The female slows briefly but doesn't stop. Again, he bashes against her. Then he reaches forward and bites her on the leg. With a loud hiss, like air being let

out of a tire, the female collapses to a stop.

Moving around to her front, the male raises his head and opens his mouth. The female pulls her head into her shell, covering her face with her huge front legs. Giving her another nip, just to make sure she won't run away, the male moves behind her and climbs onto her shell.

Balancing Act

He doesn't mate right away, but lunges up and down, bellowing like a noisy cow. He still must convince the female he's the greatest male around. When she finally lifts her shell up off the ground, they mate.

It's not easy for the male to keep his balance on top of the female, but the shape of his plastron helps. It is sunken in, or **concave**. If it were flat, he would slide off.

Male Galápagos tortoises are larger than the females. Domed males may weigh 500 to 600 pounds (227-272 kilograms) or more, while females weigh about 250 pounds (113 kilograms). Saddlebacks are smaller. Males weigh between 100 and 150 pounds (45-68 kilograms), and females only about 50 pounds (23 kilograms).

The Search for a Nest

After mating, the female leaves the male and begins her **migration** to an open nesting area in the lowlands. Several months may pass before she arrives. By then, the dry season will have begun. Because nesting areas with deep soils are scarce, females often return year after year to the same place. On some islands, the nesting areas are so small that females arriving late in the season dig up the eggs laid by earlier arrivals.

As she travels, the female's eggs begin to grow inside her body. After about two or three months, hard shell layers form around the eggs. Then the female is ready to lay. Pacing back and forth, she tastes the soil with the tip of her fleshy pink tongue, probably to sense if the ground is moist enough

Scraping the ground with her strong toenails, a female Galápagos tortoise digs a nest for her eggs.

for her eggs. When she finds the perfect spot, she begins to dig.

Digging the Nest

Resting on her shell, she scrapes the earth with her strong toenails, stroking with one foot at a time. Lifting the dirt with her back feet, she drops it on either side of the hole and pushes it away with the bottoms of her feet. If the ground is hard, she urinates to soften it. Continuing, she scrapes out clumps of mud until the hole is as deep as she can reach with her back legs—about 12 inches (30 centimeters).

Now she moves her feet to the sides of the nest and waits. Inside her body, the eggs move down their channel toward the opening in her tail. A slippery jelly, like raw egg whites, lubricates the opening like oil in an engine. It will help the hard-shelled eggs to pop out easily.

Filling the Nest

The female's tail swells, and the first egg drops out into the nest. A minute later, a second egg appears and rolls down beside the first. Large and round, about 2½ inches (6 centimeters) across, the eggs look like the white cue ball in the game of pool. Every minute or two, she lays another egg. If she is a domed female, she will lay about nine. Saddlebacks lay fewer eggs—about five.

Although the female began digging her nest in

the afternoon, she will not lay her eggs until after dark. Although she has no natural predators in Galápagos, she has inherited the instinct from her ancestors to nest under the cover of night. After laying, she spends the rest of the night covering the eggs, scooping the dirt back into the hole with the bottoms of her feet.

At last, she presses down the nest with her heavy shell, moving back and forth like a person grinding something into the ground with his foot. When she finishes, the top of the nest will be tightly packed down, encasing the eggs within a warm, moist chamber. By morning, she will be gone, migrating back to the highlands. She won't see her eggs hatch or take care of her young. They will know what to do without her.

New Life Underground

Day after day, the sun beats down on the soil, **incubating** the eggs within the nest. The warmer the temperature, the faster they incubate, although too

Taking its first peek at the world, a Galápagos tortoise slowly chips its way out of its egg.

much heat will spoil them. With warm weather, they will hatch in three or four months. With cool weather, they may take six months or more.

Early in the wet season, the tiny tortoises inside the eggs chip their way out, using an **egg tooth**, a thornlike point on the end of their nose. It may take them several days, or longer. They won't need to eat because a yolk sac bulging from

their plastron nourishes them.

When they finally emerge from their shells, the **hatchlings** scratch at the dirt with tiny, pointed claws. Gradually, they dig up and out of the nest. How do they know which way is up? Stand on an incline, like a sloping lawn, and close your eyes. Carefully take a few steps up the slope. Your body, like the tortoise's, can feel the right direction because of the pull of gravity.

Future Giants

Once free, the miniature tortoises, about the size of half an orange, scramble every which way. They have no natural enemies, except perhaps the Galápagos hawk, but they face other hazards. The islands are covered with the rubble and crevices of old lava flows. They might fall down a crack or be crushed by a falling rock.

Can you picture the tiny hatchlings, small enough to fit in the palm of your hand, growing into the giants of the Galápagos Islands? It will take a while.

In two or three years, they may be the size of half a grapefruit; in five or six years, perhaps the size of half a basketball. It will take about 25 years for them to **mature**. Even then, they may be only about half their full size.

As they grow, their shells must grow with them. The bone plates add new bone around their edges. New scute layers form under the old scutes. The scutes overlap on top of the bone plates about halfway, giving the shell strength. Look at a brick wall. Are the bricks stacked one right on top of the other? Or do they overlap so that each brick on top fits over two bricks beneath it? Try building a wall both ways with blocks. Which is easier to push over? Which is stronger?

The shell normally grows during the wet season. Each time it stops growing, shallow folds, called growth rings, form around edges of the new scutes. However, you can't tell a Galápagos tortoise's age by counting its rings. During dry years, when there is little food, it might not grow at all. Its shell would not

form growth rings, and it would be older than the number you would count. During wet years, when food is plentiful, it might grow several times, adding several rings. It would be much younger than the number you would count.

You can only tell a tortoise's age if you know when it hatched. That's why people don't know how long the Galápagos tortoise lives. To find out, you would have to find a nest, wait for the eggs to hatch, and then follow the young for more than a hundred years.

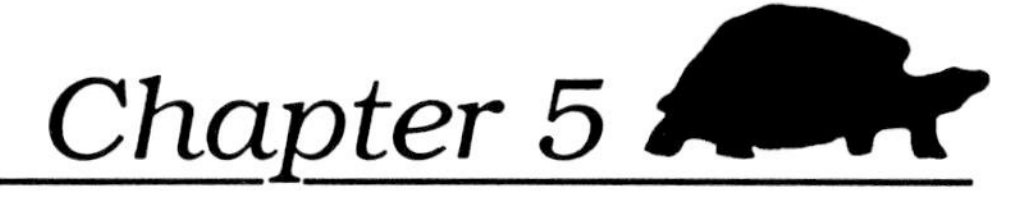

Chapter 5

Making a Comeback

The year is 1792. With a grinding crunch, a rowboat washes forward against a rocky shore. Quickly, several sailors jump out. Hoisting ropes and poles to their shoulders, they head inland in search of the Galápagos tortoise.

They hope it won't take long. Their captain is anxious to set sail, and their food supplies are low. Just one of the giant beasts would feed their crew for several days. They need many more, however, for the long ocean voyage ahead. Dozens of the tortoises could be kept alive, turned on their backs in the hull of the ship, for more than a year without food or water. That would mean a steady supply of fresh meat. The men pick up their pace.

The sailors, pirates, sealers, and whalers of the

For hundreds of years, Galápagos tortoises have been hunted for their meat and oil. Today, they are an endangered species.

17th, 18th, and 19th centuries had no idea how close they would bring the Galápagos tortoise to **extinction**. For hundreds of years, ship after ship visited the islands, taking the tortoises for food and oil. The tortoises were easy to catch. They weren't afraid of people, and they couldn't run away. Hundreds were taken at a time. By the time they were protected by law, more than 100,000 had been killed. They are now an **endangered** species.

Originally, they lived on ten different islands. Today, fewer than 15,000 remain on only six islands. On Isabela, the tortoises live on five different volcanoes. Because Isabela's volcanoes reach elevations of 3,700 to 5,600 feet (1,122 to 1,699 meters), it was difficult for people to climb them, and more tortoises survived there.

Several other tortoise populations, however, were totally wiped out. The smallest islands were the hardest hit because they were easier to cross. On the island of Pinta, only one male tortoise survived. Finally, in 1959, the Ecuadorian government estab-

lished the Galápagos Islands as a national park. Soon after, the Charles Darwin Research Station—named for the famous scientist who visited there in 1835—was built on the island of Santa Cruz.

Charles Darwin was the first scientist to notice that the animals in the Galápagos were not only different from those on the mainland, but also were unique from island to island. He believed that animals, like the tortoises, would have come from a common ancestor, but gradually **evolved** to look different as they adapted to new environments on the different islands.

The newly built research station renewed scientific interest in the Galápagos. Its mission was to conserve the unique plants and animals of the islands, including the Galápagos tortoise. It did not have an easy task.

Over the centuries, people had introduced animals—goats, pigs, cattle, donkeys, dogs, cats, and rats—that were not native to the islands. The tortoises had no defenses against them. The dogs, cats, and rats killed their young. The donkeys trampled their nests, and the pigs dug them up, eating the eggs. The goats and cattle ate the grasses and leaves the tortoises depended on for food.

To save the tortoises, scientists and park wardens dug up their nests and raised their young at the research station until they were large enough to be safe from rats, dogs, and cats. Whenever possible, the foreign animals were removed.

On the island of Española, where the smallest of the saddlebacks lived, goats had eaten every bit of food. Only 14 adult saddlebacks, their shells barely reaching two feet in length, remained. Until the goats could be removed, the two males and twelve females were taken to the research station for breeding. Any young they produced could one day be released back into their homeland. Unfortunately, with only two males, the youngsters would be closely related.

Then, in 1976, a scientist visiting the San Diego Zoo in California found a third male Española tortoise. The tortoise had been at the zoo for more than 40 years, without a female to his name. He had spent most of his days alone in a corner, avoiding the many domed tortoises in his enclosure.

Without delay, the male was returned to the Galápagos and placed with a group of his own Española females. He was home, and over the next 15 years he would produce more than 200 young. As of 1991, hundreds of young Española tortoises have been released into the wild, and their kind is making a comeback.

Today, people from all over the world visit the islands to view the beauty and uniqueness of the land, plants, and animals. They now shoot only with cameras, and collect only memories of a magical place—the Galápagos, Islands of the Tortoises.

Sources of Information about the Galápagos Tortoise

Write to:

Susan Schafer
Reptile Department
San Diego Zoo
P.O. Box 551
San Diego, CA 92112

World Wildlife Fund
Public Information Office
1250 24th Street NW
Washington, DC 20037

Show your support of programs that protect the Galápagos tortoise by writing to the following:

Program Administrator for the
Galápagos Support Program
c/o National Zoological Park
Education Building
Smithsonian Institution
Washington, DC 20008

The Director of the Charles Darwin Research Station
Estación Científica Charles Darwin
Isla Santa Cruz
Galápagos, Ecuador
South America

Call your local museum or zoo to find the name of a Turtle and Tortoise society near you.

Glossary

adapted—having changed to new conditions or surroundings in order to survive.

barren—without life

browsing—feeding on the twigs and leaves of shrubs or trees

carapace (KAIR-uh-pays)—a bony shield covering the back of the tortoise; the top part of its shell

concave—hollowed, or rounded inward like the inside of a bowl

dehydrate—to remove water from

egg tooth—a sharp point on the nose of an unhatched reptile that is used to break through the eggshell

endangered—in danger of becoming extinct, or dying out

evolved (ih-VOLVD)—gradually changed over a long period of time

extinct (ehk-STINGKT)—no longer living anywhere on earth

family—a group of related plants or animals. The Galápagos tortoise is a member of the tortoise family.

fossil—the hardened remains of plants or animals that lived long ago

grazing—feeding on grasses and other small herbs that grow close to the ground, as in a meadow

habitat (HAB-ih-tat)—the place where an animal or plant naturally lives and grows

hatchling—an animal that has recently come out of its eggshell

herbivore (HER-buh-vore)—a plant-eating animal

hierarchy (HI-uh-rahr-kee)—a system in which animals have higher and lower positions of power

incubate (INK-yuh-bayt)—to hatch eggs by keeping them warm

insulator (IN-suh-LAY-tor)—a layer of material that keeps something warm by slowing the flow of heat. A tortoise is insulated by its shell.

internal—located inside something, such as the body

invertebrate (in-VUHR-te-brate)—an animal that lacks a spinal column, or backbone

mature—to grow to an adult size and be able to produce young

metabolic rate—the speed at which a plant or animal changes its food into energy

migrate—to move from one area to another for feeding or breeding

order—a group of related plants or animals, larger than a **family**. All turtles belong to the same order.

parasite (PAR-uh-site)—an organism, usually harmful, that lives in or on another organism

plastron (PLAS-tren)—a bony shield covering the belly of the tortoise; the bottom part of its shell

predator (PREHD-uh-tuhr)—an animal that lives by killing and eating other animals

ritual—a special set of actions performed by one animal that sends a message to another animal

scientific name—a two-part Latin or Greek name given to an animal or plant based on its special features.

scutes (SKI EWTS)—the external horny plates that cover the bony pieces of a tortoise's shell

species (SPEE-sheez)—a group of animals or plants that have certain characteristics in common. Lions and tigers are two different species of cat.

subtropics—the region of the earth bordering on the tropics

terrestrial (tih-RES-tree-il)—living on land

territory—an area chosen as its own by an animal or group of animals

tropics—the region of the earth along the equator that is noted for its hot, wet climate

vertebrate (VUHR-te-brate)—an animal that has a spinal column, or backbone. Human beings and other mammals have backbones, as do fish, amphibians, reptiles, and birds.

Index 

About the Author

At her home near San Diego, Susan Schafer remembers the months she spent studying the giant tortoises on the Galápagos Islands. She is assistant curator of herpetology at the San Diego Zoo, where she shares her appreciation of amphibians and reptiles, especially tortoises, with people of all ages. She is the author of *The Komodo Dragon*, another book in the Dillon Remarkable Animals series. Ms. Schafer lives in Spring Valley, California, with her husband, a menagerie of pets, and a tiny carving of a Galápagos tortoise to remind her of the rèmarkable giant.